Shark Shock!

Tiger Shark

By Rachel Rose

Consultant: Erin McCombs
Educator, Aquarium of the Pacific

Bearport Publishing

Minneapolis, Minnesota

Credits

Cover and title page, © Serge Melesan/iStockPhoto, © sumroeng chinnapan/Shutterstock, © Memories Photos/Shutterstock; 3, © nabil refaat/Shutterstock 4, © Peter Hall/iStockPhoto; 4-5, © Rodrigo Friscione/Getty; 6-7, © imageBROKER/Alamy; 8-9, © Michael Bogner/Shutterstock; 10-11, © Terry Moore/Getty; 12-13, © Jeff Milisen/Alamy; 13, © dw__ross/Wikimedia; 14-15, © Andrey Nekrasov/Alamy; 16-17, © Media Drum World/Alamy; 18-19, © Hannes Klostermann/Alamy; 20-21, © Aaronejbull87/Shutterstock; 22, © Matt9122/Shutterstock; 22-23, © nabil refaat/Shutterstock; 24, © nabil refaat/Shutterstock.

President: Jen Jenson
Director of Product Development: Spencer Brinker
Senior Editor: Allison Juda
Associate Editor: Charly Haley
Designer: Colin O'Dea

Library of Congress Cataloging-in-Publication Data

Names: Rose, Rachel, 1968- author.
Title: Tiger shark / by Rachel Rose.
Description: Minneapolis, Minnesota : Bearport Publishing Company, [2022] | Series: Shark shock! | Includes bibliographical references and index.
Identifiers: LCCN 2021026688 (print) | LCCN 2021026689 (ebook) | ISBN 9781636915340 (library binding) | ISBN 9781636915432 (paperback) | ISBN 9781636915524 (ebook)
Subjects: LCSH: Tiger shark--Juvenile literature.
Classification: LCC QL638.95.C3 R658 2022 (print) | LCC QL638.95.C3 (ebook) | DDC 597.3/4--dc23
LC record available at https://lccn.loc.gov/2021026688
LC ebook record available at https://lccn.loc.gov/2021026689

For more information, write to Bearport Publishing, 5357 Penn Avenue South, Minneapolis, MN 55419. Printed in the United States of America.

Contents

Sneak Attack

A tiger shark cuts through the ocean at night. It spots a turtle in the dark water. Swimming slowly, it moves toward its **prey**. Then, with a sudden burst of speed, the shark clamps down on the shell with its mighty jaws. *Snap!* It's dinnertime for the powerful **predator**.

A tiger shark's strong jaw can crack right through the shell of a turtle!

Something's Fishy Here

Like all sharks, these powerful hunters are fish. They are **cold-blooded** animals that use **gills** to breathe. In fact, tiger sharks are some of the biggest fish in the ocean. They grow to be about 10–14 feet (3–4.3 m) long—about the same as a car!

Tiger sharks get their name from the stripes young sharks have on their bodies. The stripes fade as they age.

Keep On Movin'

These large fish travel through oceans all around the world. They don't stay in one place for very long. Tiger sharks **migrate** with the seasons. In colder months, they head to warm, **tropical** waters along the coasts. In warmer months, they move back to cooler waters away from land.

TIGER SHARKS AROUND THE WORLD

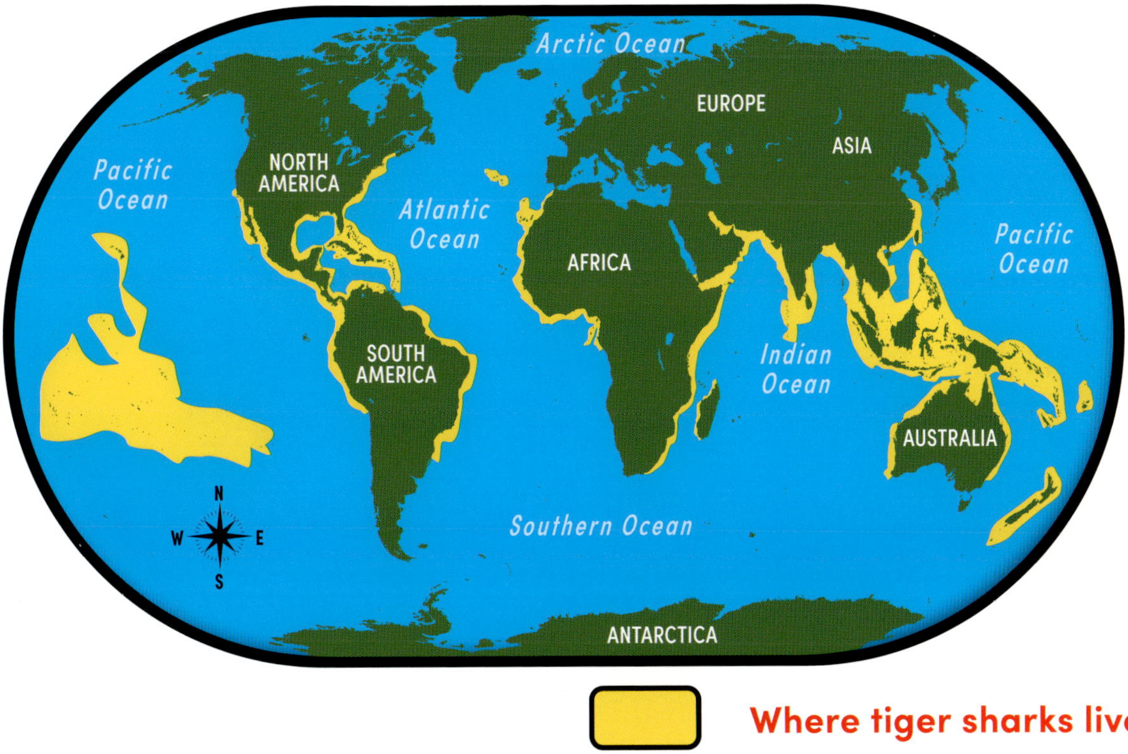

Some tiger sharks migrate over 4,000 miles (6,400 km) each year.

Hide-and-Seek

Wherever they go, tiger sharks are some of the ocean's most skilled hunters. These powerful creatures have amazing eyesight and a great sense of smell, which allows them to easily track their next meal.

They also have an extra sense that helps them hunt. Tiger sharks can feel **electrical signals** that animals give off even when the sharks can't see their prey.

Tiger sharks usually hunt at night. The sharks typically approach their prey slowly before putting on a quick burst of speed to catch the food.

Tooth Trap

Once this great hunter clamps down on prey, it's no contest. The tiger shark has powerful jaws full of incredibly sharp teeth. Like other sharks, the tiger shark has several rows of teeth. The jagged edges of the sail-shaped teeth can cut through skin, bones, and shells.

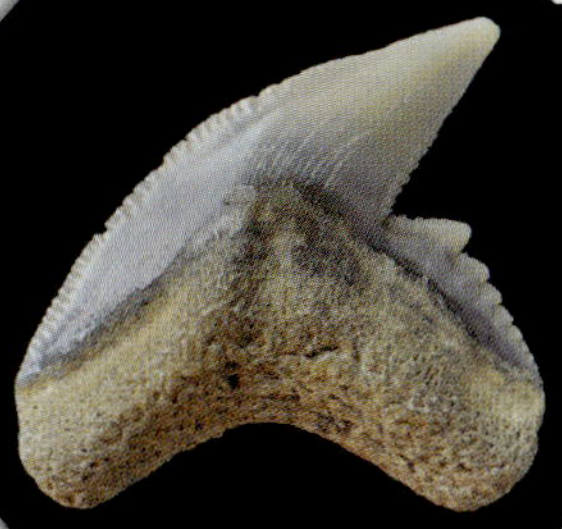

Sharks have many rows of teeth. So, when a tiger shark loses a tooth, there is another one right behind it.

I'M NOT PICKY!

What is the tiger shark's meal of choice? Anything really! Its **diet** includes other fish, seals, squid, and turtles. The tiger shark also chows down on seabirds, snakes, dolphins, and sometimes other sharks! But these feasts go beyond typical prey, too. The tiger shark has been known to eat car tires, bottles, plastic, and any other trash it can find.

Because they will eat almost anything, tiger sharks are sometimes called the garbage cans of the ocean.

The Hunter Is Hunted

Tiger sharks have very few natural predators. But sometimes, young sharks are hunted by killer whales or great white sharks.

Humans, however, are the tiger shark's biggest **threat**. People hunt them for their fins, skin, and livers, which are used in soups and medicines. Sometimes, tiger sharks are caught in fishing nets by accident.

Sharks also face **habitat** loss, pollution, and other threats from humans.

Many Pups

Tiger sharks usually live and hunt alone. But they come together when it's time to **mate**. It takes 16 months for the babies, or pups, to grow inside mother sharks. Then, mother tiger sharks give birth to about 30 pups. Occasionally, they can have as many as 80 babies at once!

A mother tiger shark has pups only once every three years.

You're on Your Own, Kid

Tiger shark mothers leave their pups right after birth. The babies are only about 20–30 inches (51–76 cm). So, at first, they are in danger from bigger predators. But within a year, the pups double in size. By the time they are around seven years old, tiger sharks are fully grown and can have pups of their own.

It is believed that tiger sharks can live for more than 40 years.

More about Tiger Sharks

Tiger sharks will eat almost anything they come across. One tiger shark was even found with a set of drums in its stomach!

Tiger sharks are the fourth-biggest sharks, after whale sharks, basking sharks, and great white sharks.

Tiger sharks can swim up to 20 miles per hour (32 kph).

Tiger sharks may lose up to 10 teeth every week. Luckily, they have plenty of teeth to spare.

Young tiger sharks have spots as well as stripes on their bodies.

The largest tiger sharks can weigh as much as a horse.

GLOSSARY

cold-blooded having blood that changes temperature depending on the temperature of the water or air around it

diet the food that a person or an animal eats

electrical signals information in the form of weak electricity surrounding some creatures

gills the body parts of a water animal that are used for breathing

habitat a place in nature where a plant or animal normally lives

mate to come together to have young

migrate to move from one place to another at a certain time of the year

predator an animal that hunts and kills other animals for food

prey animals that are hunted and eaten by other animals

threat something that could cause trouble or harm

tropical having to do with the warm areas near the middle of Earth

Index

Read More

Alderman, Christine Thomas. *Tiger Sharks (Bolt. Swimming with Sharks).* Mankato, MN: Black Rabbit Books, 2020.

Nixon, Madeline. *Tiger Shark (Sharks).* New York: AV2, 2019.

Learn More Online

1. Go to **www.factsurfer.com** or scan the QR code below.
2. Enter "**Tiger Shark**" into the search box.
3. Click on the cover of this book to see a list of websites.

About the Author

Rachel Rose lives in California. She swims in the ocean every day, and she sees plenty of seals—but she hasn't seen a shark yet!